Contents

Introduction

For a considerable time now it has been known that workplace exposure to noise beyond a certain level can lead to an irreversible deterioration in the hearing of those exposed to it.

Regulatory bodies are accordingly moving towards tougher control measures. Instruments and techniques are now available by which a worker's noise dose can be measured accurately, so it is possible to forestall the onset of noise-induced hearing loss by practicable precautionary measures. These involve consideration of design and engineering factors, and where these alone do not suffice, effective personal protective equipment is available for people to wear.

This book describes how noise affects the sense of hearing. It also examines the units of measurement for sound, and discusses the use of measurement as a feature of noise assessment and control programmes. Above all, the book reviews the range of options available to those needing to control noise and reproduces the Code of Practice against which their success is presently judged.

The book also describes the existing statutory and common law position with regard to workplace noise, and the current proposals for change in United Kingdom legislation to bring it into line with the provisions of EC Directive 86/188/EEC.

Chapter 1

Noise-induced deafness

Sound is not only one of the valued sources of information by which we interpret and appreciate our environment; it is also an agency which can injure us. Above a certain intensity the sound pressure waves bring about a deterioration in the sensitivity of parts of the inner ear and our hearing becomes slowly, insidiously, and irreversibly damaged. Other noise-induced conditions occur such as tinnitus; loudness recruitment; and psychological effects, associated with stress, fatigue, and an inability to concentrate.

The external ear

It might at first seem surprising that the ear, the organ of hearing, is damaged by sound – the stimulus it is supposed to detect. However, the human ear evolved during millions of years of relative quiet and the fact is that the current level of noise from industry and entertainment simply overloads the system.

The mechanism by which sound, in the form of atmospheric pressure waves, is converted into a signal for the brain, is highly complex. It starts with the conspicuous external part of the ear (Fig 1). This goes under a variety of names and we will use one of the more medical expressions – the pinna. The job of the pinna is to funnel sound into the next section of the ear, the auditory canal or meatus. The auditory canal is simply a tube about 2 cm long, at the end of which is a taut membrane: the eardrum. The eardrum is set into vibration by the sound wave, and this vibration is the signal which is passed on to the next section of the ear.

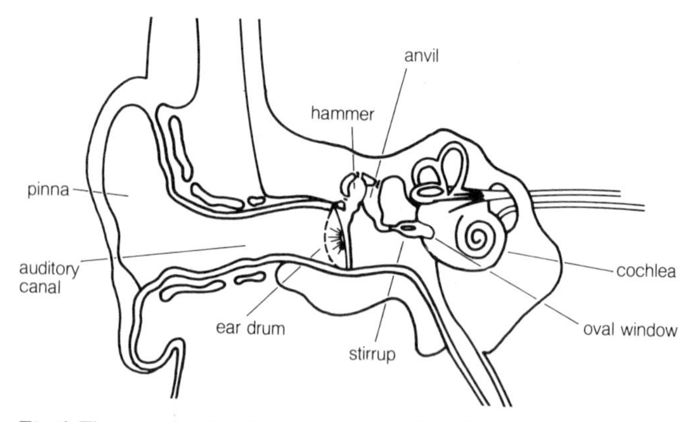

Fig 1 The ear, showing the arrangement of middle and inner ear.

The middle ear

The middle ear is a small cavity connected with the nasal cavity by the Eustachian tube. The Eustachian tube opens when we swallow so that the middle ear is kept at atmospheric pressure.

A common experience is pain in the ear when this pressure equalisation mechanism is not working. This is often the result of a cold, when mucus blocks the Eustachian tube. If the external pressure then varies, as in an aeroplane on take-off or landing, there is an uncorrected pressure differential on the two sides of the eardrum which firstly causes reduced hearing sensitivity, and then pain.

Assuming, however, that all is working well, the vibrations of the eardrums are passed on to the main components of the middle ear: the ossicles. The ossicles are three minute, connected bones: the malleus, incus and stapes, commonly known as the hammer, anvil and stirrup. These make up a small system of levers and are designed to amplify the vibrations on the eardrum. The amplified signal is transferred by the stirrup to another diaphragm, smaller than the eardrum: the oval window. The oval window is the entrance to the final section of the ear.

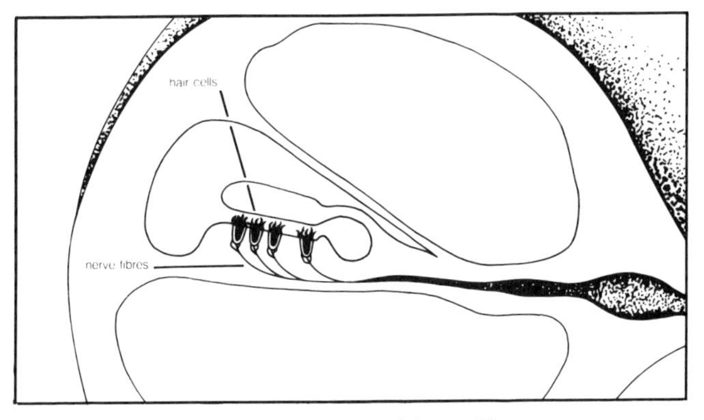

Fig 2 Diagram showing a cross-section of the cochlea

The inner ear

The inner ear is a highly complex series of tubes set in the dense part of the skull known as the labyrinth. Much of this structure is intended not to measure hearing, but to preserve the body's balance. The main hearing component, of which the oval window is part, is the cochlea. The cochlea is a coiled, fluid-filled tube, looking rather like a snail. It is divided into two parallel sections, or galleries, which are connected at the end of the cochlea. Underneath the oval window, but in the lower gallery, is another diaphragm, the round window. The round window is sufficiently flexible to allow the fluid in the cochlea to transmit the vibrations from the oval window.

It is the inner walls of the cochlea that contain the real mechanism of hearing. The cochlea is lined with about 20,000 fibres, often known as hair cells (Fig 2). These fibres are connected to the nerve endings of the auditory nerve, which leads to the brain. Thus, the hair cells, picking up the vibration from the fluid in the cochlea, generate signals in the nerve which will be interpreted by the brain as the experience of sound or, if the experience is unpleasant, noise.

The hair cells are, of course, doing more than merely detecting whether sound is present or not. They respond, with remarkable sensitivity, to variations in the loudness and frequency of sound. This enables complex signals like speech and music to be analysed for their information content, or their aesthetic value. However, the exact mechanism by which the ear assesses differences in the quality of sound, is not clear. A number of theories exist, ranging from the suggestion that different hair cells respond to different sound frequencies, to a theory based on the creation of holographic interference patterns within the inner ear. Whatever the mechanism, the hair cells are the key to hearing and to deafness caused by noise. They are the most sensitive component in a complex structure and, when the system is overloaded, are the first to break down.

Overloading

In the last two centuries or so, the human ear has been subject to bombardment with noise, quite unlike the environment in which it evolved. This is primarily because the machinery which has dominated our lives since the industrial revolution is not completely efficient. Not all of the energy in an engine goes to drive the shaft, nor all of the energy in a power press to stamping the component. Instead, energy is lost to the environment in the form of heat and noise. In addition to, and perhaps because of, our exposure to machinery noise, we have acquired the taste for exposure to high levels of sound as entertainment. Music in discotheques is played at a level well up to high industrial noise levels, and hi-fi equipment is commonly sold on the basis of its high power output level.

So what happens when the ear is exposed to high levels of noise? Firstly it introduces its own protection mechanisms. It is not always realised that the ear has some ability to protect itself, since the response is neither visible nor voluntary (as, say, closing the eyelid). The first part of the protection response is called the aural reflex. This consists of a tightening of the muscles which keep the eardrum in tension. The increased eardrum tension reduces the sensitivity of the ear, particularly to low frequency sound.

A second protective mechanism is introduced if the ear is subject to extremely loud noise. It consists of a change in the way the ossicles, the three bones in the middle ear, vibrate. Instead of directly transmitting vibration from the eardrum to the oval window, the ossicles start to rock from side to side. This greatly reduces the efficiency of the system and there is a substantial reduction in loudness.

Damage

Unfortunately, the ear's own protection mechanisms are not adequate to cope with the excesses of today's noisy environment. They fail firstly because the aural reflex takes a few milliseconds to come into operation. This means that noise resulting from impacts (such as a hammer hitting metal) which have a very short time duration, will penetrate the unprotected ear. Secondly, the in-built protection is inadequate to cope with high, sustained noise levels, day after day, week after week. That the ear is ill-equipped to handle impact noise and sustained high level noise is not surprising since these are circumstances which do not occur in nature. Rather than wait for a few million years of evolution, we are thus obliged to introduce artificial measures to protect us against this recent exposure.

The problem, as already suggested, lies with the hair cells in the cochlea. When subject to excessive vibration they become physically damaged. As a result, the strength of signal to the auditory nerve is reduced, and the victim suffers progressive deafness.

If the exposure to noise is reasonably short, and the level not too great, the resulting deafness is temporary. This is described as a temporary threshold shift, which means that the lowest level of noise which can just be heard is increased for a period of time. The higher the noise level and the longer the exposure, the greater is the threshold shift, and the longer the time for hearing to return to normal. Substantial exposure to high industrial noise can result in a temporary threshold shift which has not recovered by the time the sufferer returns to work the next day. The next period of noise exposure thus reinforces the level and duration of the deafness.

Temporary threshold shift is likely to be accompanied by the unpleasant experience of tinnitus. This is a constant buzzing or ringing in the ears, felt by many deafness victims to be a worse affliction than the loss of hearing.

Deafness

As stated, the threshold shift resulting from damage to the hair cells is temporary, provided the noise exposure is not too great, and there is a sufficient period in a quiet environment for the ear to recover. A common analogy to explain this is the damage to a field of wheat after strong wind. The wheat is compared to the hair cells and it is suggested that provided the wind is moderate, the wheat will recover in the calm weather which follows.

Farming colleagues tell me this is a poor comparison and that flattened wheat never actually recovers! Be that as it may, it does seem that the flattened hair cells can spring back into action with the result that hearing returns to normal.

However, this sequence of damage and recovery cannot go on indefinitely. If noise exposure is repeated day after day, the recovery is not complete and a degree of permanent deafness sets in. This is because the repeated stressing of the hair cells weakens them to the point where they fracture. The consequent loss of hearing sensitivity is then irreparable since the hair cell has no capacity for regeneration.

A major problem with deafness arising in this way, is that its onset is slow and insidious. Victims are very unlikely to realise that they are going deaf because on each occasion of noise exposure they lose so few of their 20,000 hair cells that they cannot perceive the difference in hearing sensitivity. In addition they may be experiencing a cycle of temporary threshold shift, in which the hearing appears to have recovered each morning, which masks the relentless progression towards deafness from which they are not recovering.

Victims will, at first, make adjustments for their deafness. They will increase the volume of the television. They will suspect that members of the family are beginning to mumble. They will compensate by unknowingly lip-reading. Everything will be heard as if through cotton wool, but so gradual is the change that they do not recognise it as a handicap until it is too late.

An important characteristic of noise-induced deafness is that the loss of hearing is not the same at all frequencies. We will look at the unit of frequency measurement, the Hertz, in the next chapter. It is sufficient to say here that the main loss of sensitivity is to sounds around 4000 Hertz. Unfortunately, this is a very important frequency for speech communication – it means that consonant sounds, such as "t" and "m" will not be heard, resulting in the feeling that sounds are muffled. In other words noise induced deafness, especially in its early stages, is a distortion of hearing rather than a uniform reduction of sensitivity to all sounds. A most serious consequence of this is that noise-induced deafness cannot be corrected by using a hearing aid. The hearing aid will merely amplify the distorted signal and will not restore the vital information content in the 4000 Hertz area which has been lost for ever.

Individual susceptibility

Like many chronic illnesses caused by workplace exposures, noise-induced deafness affects some individuals more severely than others. Thus, after exposure to the same level of noise for the same length of time, some individuals might become substantially deaf, whilst others might escape largely unscathed. As we will see later, this presents a major problem in devising noise legislation – should it protect most people, or everyone? Unfortunately, there is no way of identifying the particularly susceptible individuals until they actually start to go deaf.

The graph (Fig 3) shows an estimate of the percentage of a typical industrial population which will suffer hearing loss after exposure to noise. Again, units are defined in the next chapter. Taking some figures from the graph it will be seen that, at age 65, about 40% of those exposed to a lifetime noise exposure of 90 dB(A) will experience more than 30 dB hearing loss whilst about 10% will suffer a more serious hearing loss in excess of 50 dB.

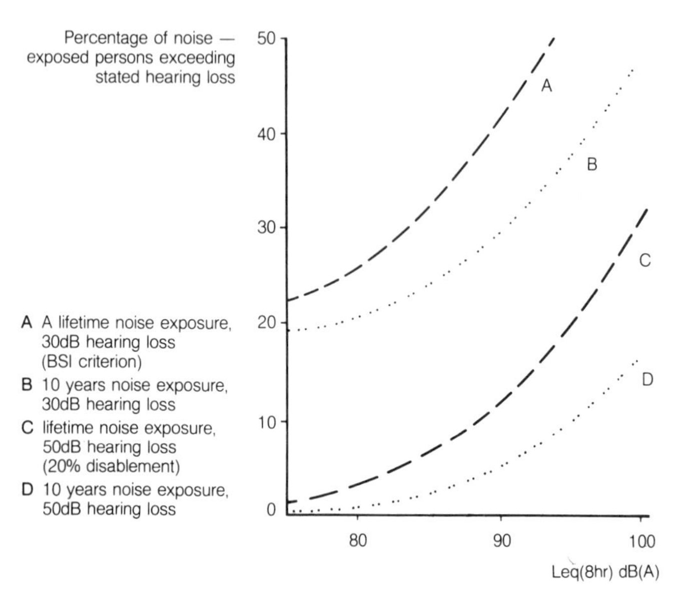

Fig 3 Hearing loss in a typical industrial population at 65 years of age (HSE estimate). Based on graph in HSC Consultative Document "Protection of Hearing at Work".

Other effects

This book is concerned primarily with the prevention of noise-induced deafness. This is a permanent, disabling affliction which isolates people from the world and excludes them from a wide range of social activity. However, it is worth noting that noise has other undesirable effects.

Tinnitus, or "ringing in the ear" has already been mentioned. It is a permanent condition in a proportion of people with noise-induced deafness. It is incessant and inescapable and cannot be corrected by surgery. Tinnitus is a major source of stress for many deafness victims.

Loudness recruitment is another common feature associated with noise induced hearing loss. It is the experience of hearing some sounds as disproportionately loud and distorted. This phenomenon is unpleasant in itself, and further reduces the ability to understand communication.

Finally, but difficult to quantify, are the psychological effects of noise. Stress, fatigue, inability to concentrate, disturbances of sleep, are all commonly associated with exposure to noise. It is therefore inevitable that work in a noisy environment will not only be hazardous and unpleasant, but also inefficient.

Chapter 2

Measurement

The principal purpose of measuring sound at work is to identify areas in which the level is injurious to those exposed to it. To arrive at this point we must know about both loudness and frequency, and develop units of measurement which can be utilised to indicate when a worker's noise dose approaches the danger level. Sophisticated instruments exist to do this, and measuring techniques also exist to assess an individual's degree of hearing sensitivity (audiometry).

Difficulties

Measuring noise presents a difficulty. It is that noise is not a measurable physical property, like heat or light, but is a subjective experience created within the brain. The dictionary definition of noise is "unwanted sound" and the "unwanted" qualification will vary from time to time and from person to person.

In practice however, sound and noise are invariably associated with a measurable physical phenomenon. That is a wave of oscillating pressure travelling through the air. This wave emanates from a vibrating object. As the object vibrates outwards, it slightly compresses the air next to it. However, air is an elastic or "springy" medium and instead of staying compressed, the air bounces back into a state of slightly reduced pressure. Whilst this is happening, the next "layer" of air is also being affected. When layer one expands, layer two will be compressed and vice versa. And so it goes on with the pressure variations being passed through the air until they fade out through dissipation and loss of energy. If we plot a graph of the pressure variations in the air, we get a wavy line showing the pressure rising and falling either side of normal atmospheric pressure (Fig 4).

On the face of it, measurement of sound or noise should now be straightforward, since our pressure wave can be fully defined with only two measurements.

These are the height or amplitude of the wave, which will indicate how loud the sound is and the distance apart of the waves, which will give the sound's frequency. In practice, there are a number of complications.

Loudness

Loudness is a function of the magnitude of pressure variations in the atmosphere as the sound wave passes through. The greater the variation, the more will the sound deflect the eardrum, and the greater will be the signal via the auditory nerve to the brain. We can therefore measure loudness in units of pressure. The problem with doing this is that the human ear is sensitive across a remarkably wide range of pressure variation.

If we were measuring pressure in Newtons per square metre, the range of human hearing would be from around 0.00002 to 20,000. In other words, we would have to use a scale with a range of about a billion units. In practice, we are simply not accustomed to coping with scales of this size, and the scale is therefore compressed by taking the logarithm to the base 10 of our unwieldy unit. The logarithm is not in fact taken directly from the pressure of the noise, but from its intensity, or amount of energy falling on a fixed area. In addition, in order to avoid awkward units, the intensity is divided by a fixed reference intensity. Since the intensity is proportional to the square of the pressure, we can now write our unit of loudness as:

$$\log_{10} \frac{p^2}{pr^2}$$

where p is the sound pressure level, and pr is the reference pressure.

The reference pressure is set at the assumed threshold of hearing (the lowest pressure to which the ear will respond) of 0.00002 Newtons/m^2.

The unit defined above is called the Bel. It is named after Alexander Graham Bell, the American inventor. Bell is most noted for the invention of the telephone but, appropriately, he also devoted a good deal of his life to the teaching of the deaf. The Bel has in fact compressed the noise scale rather too much and gives an everyday range from 0 to about 15. The scale is therefore expanded by 10 to give the common unit of noise intensity, the decibel, with a range from 0 at the hearing threshold to about 150 – a level loud enough to cause pain.

We have not yet finished with the decibel but, before discussing it further, we must consider the other characteristic of sound, its frequency.

Frequency

The frequency of sound or noise is simply the rate at which the pressure fluctuation in the atmosphere varies. Rapid variations represent a high frequency sound and the effect, in terms of

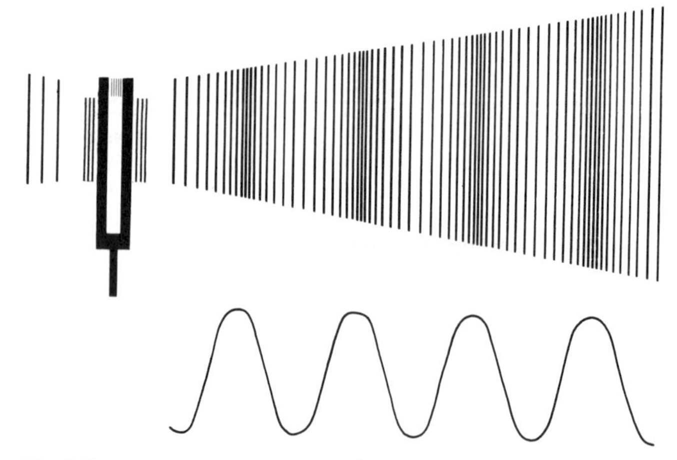

Fig 4 Diagrammatic representation of sound pressure waves.

subjective experience, is of a high pitched note. Slow variations produce low frequency sounds.

Measurement units are straightforward: the number of pressure oscillations which occur in one second are counted and expressed as cycles per second.

To complicate matters slightly, the cycle per second is now generally referred to as the Hertz, after the German scientist Heinrich Hertz who studied wave motion in electro-magnetic radiation. The abbreviation for Hertz is Hz.

Now comes the real problem – the human ear does not hear all frequencies of sound equally well. Sounds with a lower frequency than about 20Hz, and a higher frequency than about 20,000Hz, we do not hear at all. Within the sounds that we do hear, the "audible range", the sensitivity of the ear varies considerably. Maximum sensitivity is at about 3000Hz, with sensitivity falling off above and below this point.

The fact that the ear has a different sensitivity at different frequencies has an important implication. It is that the decibel cannot be used as a true measure of loudness. The reason is that two sounds could have equal decibel levels (ie they would be creating the same pressure fluctuation in the atmosphere), but if one sound had a frequency at which the ear is highly sensitive, say 3000Hz, and the other sound had a frequency at which the ear has poor sensitivity, say 500Hz, they would sound very different. Because the ear is so much better at detecting the 3000Hz sound, it would judge it to be louder than the 500Hz sound. In fact, even though the mathematical decibel levels are exactly the same, the 3000Hz sound would be judged to be about 20 decibels louder.

The A-weighted decibel

Loudness, as perceived by the human ear and brain, is a function of both the intensity and the frequency of the sound pressure wave and this means that any unit of true loudness must take both factors into account. This is achieved by taking the decibel level, as already defined, and correcting it according to the frequency of the sound. Thus, sounds at frequencies which the ear does not hear well have large correction factors, or "weighting" factors, applied so that the resulting decibel level more closely resembles the level that the ear actually hears.

A number of attempts have been made to define exactly what weighting factors should be applied to each frequency. A complication is that the factors themselves vary according to the intensity of the sound. However, for the vast majority of applications, the weighting factors are universally agreed and are usually presented as a curve – the "A"-weighting curve (Fig 5). The shape of this curve is also, of course, a plot of the variation in sensitivity of the ear with frequency. A decibel level to which the appropriate A-weighting has been applied is known as an "A-weighted decibel", abbreviated to dB(A). This unit has wide acceptance as a realistic measure of loudness. *It is the basis for noise standards and legislation throughout the world.*

Measuring dB(A)

In reviewing the units of sound and noise so far, reference has been made to "the frequency" of a sound. In practice, the sounds we experience in everyday life are an incredibly complex mixture made up of many different frequencies. For all that, however, it is always possible to unravel the sound into a series of simple pressure sine waves superimposed upon each other. Nevertheless, the business of sorting out a sound into its component frequencies, and then applying the appropriate A-weighting, involves considerable measurement and calculation. Fortunately, the increased interest in noise has coincided with the development of sophisticated integrated electronics. This means that the job of producing a combined A-weighted decibel from a complex noise can be undertaken instantly by a compact portable sound level meter. We expose the microphone of the meter to the noise, and read off the dB(A) level from a dial or a digital display.

Estimating dB(A) level

Having shown the origin of this complex unit, the dB(A), it is interesting to look at the actual levels of some common sounds. At the bottom of the range, it is theoretically possible to hear a level of 0dB(A). This is because the sound pressure level is then at the defined threshold of hearing - the reference pressure in the equation of the decibel. In practice, though, the quietest sounds that most of us can hear are likely to be more than 20dB(A). This would be the level of a quiet breeze in the countryside with no contribution from traffic.

Such an experience is now rare, and the lowest level we are likely to find in our houses, with no television on and no traffic noise, is about 35 dB(A). Once we add noise from conversation, television, etc, the level rises to perhaps 65 dB(A).

Continuing up the scale, we might find levels of 70-75 dB(A) inside a motor car. Outside the car, on the kerbside with traffic passing, the levels are around 80-85 dB(A). Heavy lorries passing by could give peaks of 90 dB(A) and above. From then on, we are into levels caused by industrial machinery. The operator of a pneumatic drill might experience 100 dB(A). Noisy factories generally have levels ranging between 90 and 100 dB(A).

The commonest source of noise levels in excess of 100 dB(A) is the aeroplane. Close proximity (25 metres) to a propellor aeroplane on take-off, gives a level typically around 120 dB(A). A jet aeroplane is closer to 140 dB(A). Levels beyond this are certainly not unknown, but are more likely to be experienced as peak rather than sustained levels. For example, a rifle shot might give a peak of 160 dB(A) at the ear of the user.

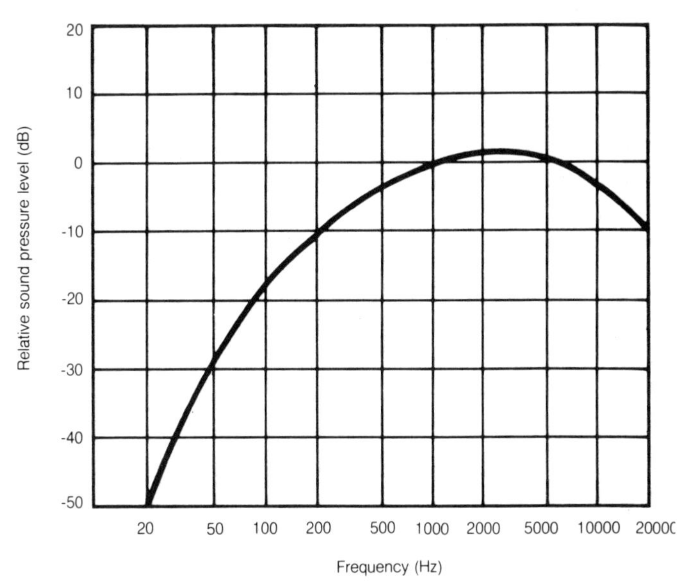

Fig 5 The A-weighting curve for sound level meters

Noise dose

Complicated though the dB(A) is, we now have to add yet another complication.

This is because we are concerned with the way noise varies with time. Exposure to noise during say, a working day, will not be to a steady, fixed dB(A) level, but to levels which are rising and falling as we move around, and use different machinery. Our real requirement is therefore not to know the dB(A) level at any particular time, but to know the dose of noise which has been received during the day.

The calculation of noise dose can be difficult. This is because the dB(A) level may be continually rising and falling, and because the logarithmic nature of the dB(A) adds a further complication in any calculation. However, the unit that emerges from the exercise is the equivalent continuous sound level, abbreviated to Leq.

The Leq is the steady level of noise which would have given the same noise dose over the measuring period, as the actual varying levels. It is also necessary to specify the period of measurement, and this is usually taken as 8 hours, to relate to a normal working day. The unit is then written Leq (8hr).

As with measuring the dB(A), modern electronics spares us much of the potential complication. A noise meter with an Leq facility continually monitors the noise level and then gives an Leq reading for whatever time it has been in use. It should be noted that an Leq (8hr) does not necessarily require 8 hours of measurement. It is simply that the noise which has been measured is averaged and spread out as though it had lasted 8 hours.

Working with the units

We have now derived some remarkably complicated units. As the book proceeds they will become more familiar, but it is worth pointing out that since the dB(A) is a logarithmic unit, it cannot be incorporated directly into arithmetic calculations. For example, if we have a machine producing a noise level of 90 dB(A), and place a similar machine beside it, the resulting level is not 180 dB(A). This is because doubling a number causes its logarithm to increase by 0.3. Doubling sound intensity therefore results in a rise of 0.3 Bels or 3 decibels. The two 90 dB(A) machines therefore will only produce a new noise level of 93 dB(A).

The 3 dB(A) doubling criterion can be useful in assessing Leq or noise dose. For example, one way of experiencing an Leq (8hr) of 90 dB(A) (an important standard in the United Kingdom, as we will see in the next chapter), is to be exposed to a level of 90 dB(A) for 8 hours. However, we could receive the same noise dose through exposure to twice the noise level for half the time. In other words to 93 dB(A) for four hours. This enables us to build up a table of noise levels and exposure times, all of which meet a given standard. For example:

90 dB(A) Leq (8hrs)
90 dB(A) for 8 hours
93 dB(A) for 4 hours
96 dB(A) for 2 hours
99 dB(A) for 1 hour
102 dB(A) for 30 minutes
105 dB(A) for 15 minutes

Audiometry

In this chapter we have so far been concerned with measurement of the characteristics of the sound or noise emanating from a source. Before concluding the review of measurement we should consider another measurable factor – the hearing sensitivity of an individual.

The measurement of hearing sensitivity or acuity is called audiometry and consists of playing pure tone sounds (ie sound of a single frequency) through headphones worn by the subject. Each ear is tested in turn and the lowest level at which the sound can be heard is recorded. The test is repeated at a number of pure tones, usually 125, 250, 500, 1000, 1500, 2000, 3000, 4000, 6000 and 8000Hz, and the graph of sensitivity at each of these frequencies is known as an audiogram. The audiogram is generally presented as the hearing loss, in decibels, at each frequency as compared with normal hearing.

As with the measurement of noise, audiometry is greatly simplified by modern electronics and instrumentation. Once set up, many audiometers will automatically plot an audiogram, the subject having merely to press a button when he or she can hear the sounds. However, audiometric testing nevertheless requires trained supervision and proper surroundings. It is also essential to ensure that the person being tested has had an appropriate period of time in a quiet environment to ensure that he or she is not suffering from any temporary threshold shift.

Chapter 3

The law

Limited specific statutory regulations exist at present on the subject of workplace noise in the United Kingdom, but changes are imminent to bring United Kingdom legislation in line with the provisions of the European Directive on The Protection of Workers from the Risks Related to Exposure to Noise at Work 86/188/EEC.

This chapter is based upon the consultative document setting out the Health and Safety Commission's proposals for change in the belief that, whilst there may be some changes in detail, the main thrust of the new regulations will be as described here because they reflect the Articles in the EC directive with which the United Kingdom has to comply.

Until such time as the noise regulations come into operation, the Code of Practice (reproduced on pages 69–79) and the Health and Safety at Work, etc Act 1974 (ss. 2, 3, and 7 in particular) govern the statutory duties all employers have in relation to the control of the exposure of their employees and others to harmful levels of noise from their activities at work.

Legislative challenge

In Chapter 1 it was seen that the hazard of noise must be expressed in terms of probabilities. In other words, at any given noise level a certain percentage of the exposed population is at risk. Hazards of this sort present a major challenge for legislators because they must decide how many people the law should protect.

At first the answer might seem obvious: the law should protect everyone. However, if this was achieved by setting a statutory maximum noise level at which no one would be affected, that level would be so low as to be virtually unattainable by most industries. Framing legislation therefore involves a compromise between reducing the risk and setting a realistic standard.

Statute law

It might seem surprising that, although the hazard of noise has been recognised since the 1960s, there is still very little statute law in the United Kingdom which deals directly with the subject. There are five exceptions:

(a) The Woodworking Machines Regulations 1974, require that noise is reduced to the greatest extent which is reasonably practicable, and that ear protectors are provided and used when people are likely to be exposed at or above 90dB(A) Leq (8hr).

(b) The Agriculture (Tractor Cabs) Regulations 1974 set a maximum level of 90dB(A) in the criteria for the approval of safety cabs.

(c) The Offshore Installations (Construction and Survey) Regulations 1974, and the Offshore Installations (Operational Safety, Health and Welfare) Regulations 1976, require the insulation of equipment capable of causing injurious noise, and provision of suitable protective equipment, including ear protection, for those at risk.

There are also regulations covering merchant shipping which deal with the noise hazard.

Beyond these exceptions, the subject of noise does not appear specifically in any statute law concerned directly with the workplace. Under the Control of Pollution Act 1974, however, a local authority is enabled to take summary proceedings in respect of a noise which is established as a nuisance. Workplace noise is one of the commoner sources of noise nuisance complaints. Employers seeking to mitigate noise inside their premises for the benefit of employees could possibly, by such action, create a noise nuisance for neighbours, eg by moving a noisy machine to an outside location. Having said that there are few specific statutory references to noise, this does not mean that there is not a general statutory requirement to reduce noise levels overall, and reduce the exposure of employees to noise levels likely to be harmful to their health.

S. 2 of the Health and Safety at Work, etc Act 1974 reads "It shall be the duty of every employer to ensure, so far as is reasonably practicable, the health, safety and welfare at work of all his employees". There is no doubt whatsoever that the hazard of noise has to be dealt with as part of this broad based duty placed upon the shoulders of employers by the 1974 Act. The Director General of the Health and Safety Executive is on record as saying in 1984 that the 1972 "Code of Practice for Reducing the Exposure of Employed Persons to Noise" (reproduced on pages 69–79) will be used by inspectors in the enforcement of employers' duties under the 1974 Act. The number of notices issued by inspectors since 1984 has increased significantly in line with the HSE Director General's declaration on policy. The Code, which is *not* an Approved Code of Practice made under the provisions of s. 16 of the Health and Safety At Work, etc Act 1974, is nevertheless one which indicates what criteria are being used by the HSE to indicate what they believe to be reasonably practicable standards to achieve. The 1972 Code sets a limit of 90dB(A) Leq (8hr) above which employees should not be exposed. At this juncture, it is appropriate to consider what has happened since the combination of the 1972 Code of Practice and the 1974 Act first established the guidelines which it was expected industry would follow to deal with the hazards to employees from noise.

Proposals for new legislation

In 1981 the Health and Safety Commission published a document containing its proposals for new noise regulations. These were extensive and received generally encouraging comment from those concerned. In the midst of the consideration of the comments made on the 1981 Consultative Document there was a proposal from the European Commission (CEC) for a European Community directive requiring Member States to harmonise their basic legislation on the protection of their workforces from noise at the workplace. In view of this proposal, published in 1982, it was clearly thought to be inappropriate that the United Kingdom should continue developing its own proposals for noise control when the European Community was also undertaking a similar task. The plans for a Community directive proved controversial mainly because of the proposal to set a single action level/limit of 85dB(A). After discussion by the European Parliament and the Council of Ministers, the proposal for the directive was amended to include two action levels – 85dB(A) and 90dB(A). The agreed Directive (86/188/EEC) was finally adopted in May 1986 and has to be implemented in most Member States (including the United Kingdom) by 1.1.90.

The Health and Safety Commission of the United Kingdom has published a further set of proposals for legislative changes in a Consultative Document: "Prevention of damage to hearing from noise at work – Draft proposals for Regulations and Guidance" HMSO. The period for making comments on these proposals, made in the light of the contents of the directive, ended on 30.6.88, and the final stages in the preparation of the United Kingdom's regulations are proceeding. The intimation has been given that the Government will make regulations ahead of the deadline of 1.1.90 to allow industry time to assimilate what needs to be done. This edition has been written on the basis of the reasonably safe assumption that the latest proposals, which follow the directive, will not differ significantly from those in the second consultative document, the contents of which are reviewed below.

Proposed new regulations

The duties under the new regulations will place responsibilities for compliance on employers, the self-employed, employees and the manufacturers of articles for use at work. The main requirements are as follows:

Reduction of risk of hearing damage

Apart from any specific steps referred to later, *every employer is under a statutory duty to reduce the risk of damage to the hearing of his or her employees from exposure to noise.* The reduction has to be to the lowest reasonably practicable level. In view of the wording in s. 40 of the Health and Safety at Work, etc Act 1974, it is the employer who must be in a position to prove that it was not reasonably practicable to do more than was in fact done to satisfy the duty. It should be noted, however, that where a prosecution is pursued, it is normally the prosecutor who has to prove beyond reasonable doubt that what is alleged is actually the case. Where the onus of proof lies with the accused, the tribunal only has to be satisfied on the balance of probabilities.

To determine what is reasonably practicable the employer may weigh the potential risk created by the noise exposure against the cost (in money, time and trouble) involved in reducing it. Measures must be taken unless their cost is disproportionately high in relation to the benefit they yield.

Assessment of the potential risk should take account of modern information about the relationship between noise exposure and the damage to hearing. Guidance is available from official sources (HSC or HSE) and from some of the guidance in British Standard 5330:1976 "Method of test for estimating the risk of hearing handicap due to noise exposure".

Reduction of noise exposure

As well as the general duty described above to reduce the risk of damage to hearing, the employer has a much more specific duty. This is to reduce, so far as is reasonably practicable, the exposure of employees to noise, *other than by the provision of personal ear protectors.*

This duty arises when any employees are likely to be exposed to a daily personal noise exposure of 90dB(A) or more, or to a peak sound pressure of 200 pascals. It should be noted that the significant qualification of the duty lies in the use of the phrase "are likely to be exposed". They *do not have to be exposed* to such levels for the duty to arise.

The duty also emphasises that there is a statutory embargo on the indiscriminate provision of personal protection *until* other more reliable engineering controls have been applied, and have been found to be inadequate by themselves to bring the exposure level down to the appropriate level.

Ear protection

The employer's duty to provide suitable and efficient personal ear protectors is a two-stage affair:

(a) when any employee is *likely to be exposed* at or above 85dB(A), the employer must take *all reasonable steps* to ensure that the employee is provided, at his or her request, with ear protectors. There is at this stage no element of compulsion upon the employee.

(b) when any employee is *likely to be exposed* at or above 90dB(A), or to a peak sound pressure of 200 pascals, the employer must take *all reasonable steps* to ensure the employee is provided with suitable ear protectors which, when properly worn, can reasonably be expected to keep the risks to hearing to below that arising from that level. At this stage, employees have a duty to take *all reasonable steps* "fully and properly to use personal ear protectors" provided by their employer, and any other protective measures provided by him or her.

Assessment of exposure

It will have become apparent by now that before an employer can comply with all the duties referred to so far, he or she has to discover precisely what the nature and extent of the noise hazard is in the premises over which he or she exercises control. This is in fact another, and some would say the most important, statutory duty which the new regulations impose. Whenever any employee is likely to be exposed at or above 85dB(A), or to a peak sound pressure of 200 pascals, the employer must make adequate arrangements for a competent person to make a noise assessment. The two purposes of the assessment are:

(a) to identify which employees are exposed at or above the levels of noise exposure mentioned and
(b) to provide the employer with the information he or she needs to comply with the other duties concerned with protecting those identified in (a).

An assessment will have to be reviewed at once, if there is any reason to suspect its validity, or, if there has been any significant change in the work it covered.

The first assessment will have to be recorded and kept for as long as any employee covered by it still works for the employer concerned. Any review of an earlier assessment will also have to be recorded and the record kept.

Maintenance and use of equipment

Once the assessment has indicated where the risks for the noise hazard exist, the employer has to take *all reasonable steps* to ensure that anything he or she has provided in compliance with the regulations is properly used, and kept in "an efficient state, in efficient working order and in good repair".

Employees' rights and duties

Employees have a right to request suitable and efficient ear protectors when they are likely to be exposed at a daily personal noise exposure of 85dB(A) or a peak sound pressure of 200 pascals.

Employees must take all reasonable steps "fully and properly to use" personal ear protectors or other protective measures when they are provided by an employer. This, it will be remembered, is where employees are likely to be exposed at or above 90dB(A) or a peak sound pressure of 200 pascals.

Employees will recognise when this duty exists because *ear protection zones* will, so far as is reasonably practicable, be demarcated and identified by means of suitable notices indicating the need to wear ear protectors while in the zones.

Along with other key words and phrases, an "ear protection zone" is defined. It means "any premises or part of premises where any employee is likely to be exposed at or above the second action level or peak action level". The phrases "second action level" and "peak action level" are also defined. Second action level means a daily personal noise exposure of 90dB(A). Peak action level means a peak sound pressure of 200 pascals.

The point at which employees may request their employer to provide them with suitable and efficient ear protectors is 85dB(A) and this is referred to as the first action level. Whilst on the subject of definitions, there is one other which is highly relevant to the individual employee. This is the employee's "daily personal noise exposure". This means the daily personal noise exposure of an employee ascertained in accordance with Part I of the schedule to the regulations, but taking no account of the effect of any personal ear protector used.

Provision of information to employees

Perhaps the most important single requirement which will guarantee the success, or otherwise, of the regulations is the one requiring the employer to provide employees with adequate information, instruction and training.

The duty arises when an employee is likely to be exposed at or above 85dB(A), or to a peak sound pressure of 200 pascals. The duty is to describe the risk of damage to the employees' hearing which an exposure to noise may bring about. Further, it is to tell employees how they can minimise that risk and, in particular, the steps they can, and should, take to obtain personal ear protectors, and fit them properly, before entering demarcated ear protection zones.

It should be pointed out that the employer's duty to employees is a personal one towards each and every employee likely to be exposed at or above 85dB(A) or a peak sound pressure of 200 pascals.

Miscellaneous provisions

A number of other provisions appear in the proposed regulations which, whilst not central to the protection of the hearing of employees exposed to noise, are necessary for the regulations' smooth functioning.

(a) An employer is under the same duty in respect of others at work (eg: employees of contractors) as he or she is towards his or her own employees.
(b) The regulations will not apply to those on board aircraft and shipping where standards are generally regulated under international agreements.
(c) The Health and Safety Executive is given powers to exempt from certain requirements under certain circumstances, if it is satisfied that the health and safety of persons likely to be affected by the exemption will not be prejudiced in consequence of it.
(d) The requirements relating to noise in the Woodworking Machines Regulations 1974 are revoked and the new regulations will apply to woodworking machines in the same way as any other.

European directive

The EC directive on the Protection of Workers from the Risks Related to Exposure to Noise at Work (86/188/EEC) was adopted on 12.5.86.

It lays down in its 14 Articles the principles to be followed in the European Community, and requires that the Member States (with the exception of Greece and Portugal) bring about the necessary changes to their domestic legislation by 1.1.90. In the cases of Greece and Portugal the relevant date is 1.1.91. In the light of progress made in scientific knowledge and technology, as well as experience gained in the implementation of the directive by Member States, there will be a re-examination of the directive before 1.1.94.

The Council, on that occasion, will try to lay down indications for measuring noise which are more precise than the ones currently given in Annex 1 of the directive. Article 7 of the directive stipulates that workers exposed to a level of 85dB(A) should be able to have their hearing checked by a doctor and, if judged necessary by the doctor, by a specialist. The way the check is carried out will follow the national law and practice of Member States.

The United Kingdom view is that medical checks, such as described above, are available through the National Health Service. The question of audiometry is not part of the statutory regime as presently envisaged, but is a matter for individual employers to determine for themselves.

The costs and benefits

Governments require economic assessments to be made when new legislation is proposed. The 1987 Consultative Document is no exception, and puts forward some facts and figures before concluding that the overall costs cannot be considered disproportionate to the substantial benefits associated with the new proposals for changes in the law controlling workplace noise.

Initial assessments will cost about £8.5 million, and the present value of the future recurrent cost over 40 years is estimated to be of the order of £21 million.

The cost of a personal protection programme, plus the cost of providing information and training, could now be, on average, about £12 per worker per year. It is thought that some 620,000 additional workers could be involved in expenditure under the new regulations. This could cost £7 million in the first year, and £127 million over a 40 year period at present day values.

It is assumed that it will be reasonably practicable to reduce noise at source significantly, initially in 10% of cases, rising to perhaps 50% after 40 years, at an average cost of £500 – £1000 per exposed worker. This would cost £30 – £60 million in the first year.

Benefits will accrue because fewer workers will sustain noise-induced hearing loss. If this is looked at in terms of the amount by which court compensation awards are reduced, because of the smaller number of claims, the estimated saving will be £171 – 218 million. Other savings arise from some reduction in loss of future output potential, but this is not quantifiable. On the other hand, NHS costs involved in the provision of hearing aids, a figure which is quantifiable, are likely to be £4 million.

Common law

Claims for noise-induced hearing loss attributable to noise exposure at work are dealt with in the courts in the same way as other claims for personal injuries. Persons suffering the hearing loss must establish that their employer was in breach of the duty of care towards them to take reasonable care of their hearing. Only foreseeable hazards can be the subject of negligence claims, so the first thing a plaintiff has to do is to establish that there is a risk to his or her hearing from exposure to too much noise at an employer's premises for too long a period.

The courts now accept that, since the publication of an official guidance booklet "Noise and the Worker" in 1963, employers have been put on notice of the risk of injury to hearing where exposure to noise is established. It is unlikely now that the noise hazard will be disputed, so it is then a matter of the plaintiff establishing that he or she has been exposed to excessive noise, and that this exposure has been for a long enough period to cause hearing loss.

The 1963 booklet's section on "The Danger Levels of Noise" included a table showing a maximum sound level in dB(A) and a corresponding exposure duration in hours per day. The figures were for an 8 hour exposure 90dB(A); for 4 hours 93dB(A); for 2 hours 96dB(A); for 1 hour 99dB(A); for 1/2 hour 102dB(A); and for 1/4 hour 105dB(A). Later on in the section appears the statement:

For the present it is recommended that such discrete impulses should be first assessed using an ordinary sound level meter set to the fast response, and *if the meter needle passes 90dB(A) the noise should be regarded as potentially hazardous* and should be further assessed by a specialist capable of interpreting the latest scientific literature on the subject.

In theory, both exposure to a hazardous noise level and exposure to such a level for a period long enough to cause hearing loss have to be established by the plaintiff for him or her to succeed. In practice, however, it appears that, where the level of exposure is established as potentially hazardous, the duration tends not to be a critical issue and the plaintiff succeeds, almost regardless of the period of exposure.

At this stage, what has to be reviewed are the steps, if any, taken by the employer to safeguard his or her employees' hearing. What did the employer provide either in the way of enclosures or accoustic measures, or personal protective equipment to mitigate the hazard? The word "provide" is interpreted narrowly. The employer really has to have the protective equipment available at the point where the hazard is encountered. He or she also has to inform, instruct and train those exposed to the hazard in what they should do to protect themselves, and a disciplinary regime should exist to enforce this.

The final matter which the plaintiff has to tackle is to show that the loss of hearing he or she suffered came about as a result of the defendant employer's negligence. This is a matter of medical history and of investigating whether the plaintiff has been exposed to other sources of injurious noise which could explain the condition. These could include gun shots from shooting sports, or explosions during the course of hostilities, etc. If the plaintiff can establish that no such exposures were suffered, he or she will succeed, on the balance of probabilities, in establishing the defendant employer's negligence and be awarded damages.

Damages

Damages awarded to successful plaintiffs in industrial deafness cases vary according to a number of factors: age; degree of hearing loss; life-style; existence of accompanying disabilities such as tinnitus; and, sometimes, loss of earnings.

The following cases illustrate the wide range in the amount of damages awarded:

(a) 1979, moderately severe deafness and tinnitus, £7,500;
(b) 1981, severe deafness and tinnitus, loss of earnings, £15,000;
(c) 1982, deafness and loss of amenity cases, £2,500 – £6,500.

An estimate from a solicitor specialising in deafness cases puts the worth of claims actually reaching the courts as ranging from £2,500 – £15,000.

Increasingly, employers, through their insurers, are party to an agreement with the relevant Trades Union to which their employees belong. Such agreements, known as Scheme Agreements, have a sliding scale of compensation varying from a £300 award for a 60 year-old man with a hearing loss of 10-14dB up to £11,000 for a 40 year-old man with a hearing loss of up to 96dB.

Contributory negligence

Employees have both statutory and common law duties to protect themselves. If they fail to fulfil their statutory duty they commit a criminal offence and the courts can fine them. If they fail to fulfil their common law duty to protect themselves, there will be a finding of contributory negligence against them and the damages awarded to them will be reduced to an extent commensurate with their own negligence. This is a matter determined by the judge after he or she has considered all the circumstances of the particular case.

Often the issue of contributory negligence turns upon the question of the condoning of a bad practice by an employer. If, for example, a foreman fitter knows that one of his staff goes regularly into a noisy compressor house without ear protectors, and the foreman has taken no steps to enforce the use of the protective equipment by the fitter, the latter will be able to say the failure to use the protectors was condoned by the foreman. The employers, because they are vicariously liable for their foreman's act of condoning the fitter's failure to use hearing protection, will not succeed in convincing the court that the fitter was negligent. In the event, however, the full damages would probably not be awarded because the fitter would almost certainly be unable to deny, during cross-examination, that he or she was completely unaware of the existence of the risks from the hazard of noise, and that ear protectors should have been worn.

When the demarcation of ear protection zones becomes obligatory, it will be more difficult for employees to deny that they were unaware of the hazards to which their attention is constantly being drawn by notices. Nevertheless, much will still depend upon the attitude of the employer, especially at the workplace itself where any condoning of bad practices begins with shopfloor supervision.

Chapter 4

Noise control

A number of options exist to control noise at source. Efficient routine maintenance should eliminate sources such as loose machine parts, and air and steam leaks. Substitution, re-design, and re-siting may prove to be effective measures. Further control methods include the use of vibration isolators, insulation, absorption, silencers and sound havens.

When noise is excessive, the ideal response is to reduce it. Provided it stays reduced, the problems of providing and enforcing the wearing of ear protection, hearing conservation zones, and so on disappear.

However, noise reduction can be expensive and complicated. It is sometimes impossible. This chapter will review the main options available, without going into the depth of technical detail which is the province of the specialist publications given in the bibliography.

The survey

It is not uncommon to find that the introduction of expensive noise control measures comes as a disappointment to all concerned because the noise reduction is much less than was anticipated. The reason for this is that one particular noise source has been tackled, but another one now dominates. It is thus essential that before any noise control is introduced, a proper noise survey is undertaken. The survey should identify all significant noise sources and should be used to predict the effect on the overall level of noise reduction alternatives.

Basic measures

Another action to undertake at an early stage is to ensure that all readily controlled noises are eliminated. Major offenders here are air and steam leaks. These can produce a high frequency background noise to which everyone becomes resigned, but which can be removed by simple engineering measures, backed up by adequate routine maintenance.

A further common source is the rattling and drumming of machine parts which are not securely fastened. For example, bolted sections of machine guards can work loose and then set up an incessant vibration which becomes one of the major causes of workplace noise. The solution is obvious, but until such basic engineering measures are undertaken, the devising of sophisticated noise control devices is pointless.

Substitution

If workplace noise arises primarily from a small number of identifiable sources, say noisy machines, then an ideal solution would be change the machines for quieter ones. Noise control technology has advanced considerably in recent years and equipment manufacturers have been under considerable pressure to minimise noise output. It is therefore likely that a less noisy version is now available and that the supplier will be able to specify the noise output which can be expected. If such data is not provided as a matter of course, it should certainly be requested.

If substitution could theoretically solve a noise problem, the overriding objection is likely to be the cost.

The cost of replacing high capital value items may rule out this option, but it is worth ensuring that the cost analysis has taken all factors into account. These might include, for example, the possible increased output from new, more efficient, machines. Also included should be the savings in the hearing protection programme (ear muffs, noise zones, audiometry, etc) which might be avoidable altogether if the noise level is reduced.

Re-design

It may not be necessary to replace a complex machine completely if a modification to its design can reduce noise. This is a specialist subject, but possible components for re-design might be the bearings or gears of rotating machinery. Bearing noise might be reduced by improved lubrication or flexible mountings. Gear re-design might include different tooth formation, or use of plastic components in the gear train.

Re-siting

An attractive approach to noise control in some locations is not to silence the machines, but to move them to where their noise is less of a problem. This might, for example, entail re-routing pipework so that components which are noisy but require only limited access, such as pumps, are located remotely from the main operating area.

An alternative might be to group noisy equipment together, accepting the fact that this will result in an area or room requiring the use of hearing protectors, but enabling other production areas to be free of such restrictions.

This has been a successful approach with equipment such as tabletting machines which can be the only noisy section in an otherwise quiet production line. By creating a separate tabletting room, the use of protective clothing is restricted to the one area, and job rotation can reduce the time which any individual must spend in the area.

Vibration isolation

If a power driven machine stands, or is fixed, directly on a hard surface, then it is likely that energy will be transmitted through the mounting points into the surrounding structure. The consequent vibration of that structure, which may of course be the floor, can be responsible for the majority of the noise arising from the machine. The solution is to isolate the machinery by inserting vibration isolators between the mounting points and the operating surface.

Vibration isolators are generally specified according to their static deflection. This is the distance which the isolator will yield under the load of the machine. The manufacturer will generally advise on the appropriate isolator for a particular machine. It is essential to seek and follow this advice since much the best results will be obtained from isolators which are accurately specified. Indeed, installation of the wrong isolators can introduce new vibration patterns which make the situation worse.

The simplest vibration isolators consist of pads of resilient material such as cork or felt. However, these do not give good isolation over a wide range of frequencies. They can also deteriorate with age, or on exposure to water and oil. For more versatile and durable application, the majority of isolators are of rubber-in-shear type consisting of a rubber body with a metal top and base. When these are specified according to their deflection the "dynamic" deflection should be known because rubber is less yielding when it is actually vibrating than when it is static.

The main remaining type of vibration isolator is the steel spring mount. These give the greatest static deflection and therefore isolate the lowest frequencies of vibration. To prevent high frequency components being transmitted through the coils of the spring, there is usually a rubber pad between the spring and the body of the isolator.

A possible consideration in installing vibration isolators is the inherent strength of the machine frame. If the frame relies for its stiffness on the support from the floor, this will be lost when the isolator is installed. It might then be necessary to mount the machine on a new frame or bed, and then to install this on the vibration isolators.

A final factor to consider is that the machine might be transmitting vibration through routes other than its base. These might include connecting pipes and services, etc. It will then be necessary to introduce flexible connectors such as rubber hose to ensure that there is no vibration bridging path left from the machine. This is most important since the value of isolators can be eliminated if bridging paths are left. Manufacturers will advise on correct specifications.

Insulation

Perhaps the most obvious approach to noise reduction is to put a box around the source in the hope that this will enclose troublesome noise. This technique of noise insulation is indeed appropriate in many circumstances but it raises a number of issues of specification and design.

The first consideration is – of what material should the box be made? There is much confusion about what constitutes a good noise insulator. This is partly because materials such as polystyrene tiles which are installed to reduce sound reflection in rooms, are also assumed to prevent noise transmission. They do not. Materials which have good insulation properties are, in the main, those which have a high density. In addition, the material should ideally have a low stiffness.

These considerations invariably result in a compromise. The ideal high density/low stiffness material would perhaps be lead.

However, cost notwithstanding, lead would not be a sensible material from which to build an industrial enclosure. Brick, though not very high density, gives good insulation in the form of brick walls of usual thickness. But brick might not be practical for fabricating an in-plant enclosure. Steel, a good construction material of high density, also has a high stiffness which reduces its insulation effectiveness. It is thus common to find that a compromise consists of composite materials which give a reasonable combination of density (or mass), whilst reducing the effects of high stiffness.

For example, steel may be used which has had a damping layer of mastic applied to one side to reduce the effect of stiffness. It is also possible to find combinations of steel and plasterboard, mineral wool or lead.

Having selected a suitable material, it is necessary to design the enclosure. This is likely to be difficult because the enclosure cannot usually be a close-fitting, completely sealed box. Services must be fed into and taken out of the enclosure, as must the product. In addition, a good acoustic insulator will probably be a good heat insulator, and ventilation must be introduced which does not destroy the acoustic properties. It may also be necessary to fit doors and windows into the enclosure to permit viewing, and allow access for maintenance.

These considerations demand detailed, specialist design. It is disappointing to find that a "home made" enclosure produces nowhere near the hoped for noise reduction because insufficient attention has been given to the selection of materials, or to constructing an enclosure which has no gaps or holes through which noise could escape.

Absorption

When sound strikes a surface some of it is absorbed, some is transmitted through the material, and some is reflected. We have looked at the characteristics of good insulation materials, but in many circumstances it is desirable to minimise the sound reflection. Typical examples are the lining of a ceiling to reduce reverberant effects, and the internal lining of acoustic enclosures.

This latter feature is desirable because if sound reflection is reduced inside the enclosure, less sound will impinge on its surrounding walls to be imparted to the outside. As discussed, it is important to differentiate between the need for insulation and absorption. Insulation is required if noise is to be prevented from penetrating an occupied area from outside, or from an enclosure. Absorption is appropriate if the noise source is within a certain area and it is desired to reduce its reflection from surrounding surfaces.

Absorptive materials do not have the same high density requirements as insulators. They are usually materials with a porous surface structure such as mineral and glass fibre, and polyurethane foams. If the absorptive surface must be more durable than these materials, then perforated metal sheet is often fitted. A layer of thin, impervious plastic may also be included to prevent the ingress of oils, etc.

In some cases, where high levels of absorption are required, it is possible to space absorptive panels so that reflected sound strikes another panel and is progressively reduced. This is normally achieved by hanging panels vertically from the ceiling.

Silencers

Devices such as fans, which push air along a system of pipes or ducts frequently benefit from the installation of silencers. The objective is to reduce noise transmitted through the system, whilst retaining a reasonably free flow of air.

Most silencing of this sort uses a system of progressive absorption of the noise. This can often be achieved simply by lining the inside of the duct with suitable absorption material such as mineral wool. When silencing of low frequency noise is required, it is necessary to "split" the duct into sections, each of which has a width much smaller than its height. Splitter silencers achieve this with baffles of sound absorption material.

When silencing is required primarily at a specific frequency, it may be appropriate to use reactive silencers. These operate by forcing the air to oscillate through narrow constrictions in such a way that the sound waves coincide and cancel each other out. Reactive silencers can be tuned to the frequency to be controlled and are not effective at other frequencies. They are therefore suitable for constant speed machines generating noise at a discrete frequency.

Sound havens

If reducing machinery noise is impossible, or prohibitively expensive, it may be worth creating a quiet enclosure or haven for employees to occupy for at least part of their working day. The principles of construction of a haven are essentially those for an acoustic enclosure, except that their function is to keep noise out, rather than to keep it in.

The sound haven approach can sometimes be taken to elaborate lengths by connecting instruments and controls within the enclosure. It may then be possible for the majority of process activities to be undertaken in a quiet environment with relatively little need to enter the noisy plant areas.

Of necessity this has been a rather basic review of a technically complex subject. The main methods of noise reduction have been described, and emphasis has been placed on the importance of undertaking the simpler, cheaper options first. However, the detailed design and construction work requires specialist input, and is dealt with in the textbooks on acoustic treatment.

Chapter 5

Hearing protection

When noise can not, within the constraints of reasonable practicability, be reduced at source, the individual worker's exposure to it may be reduced by his or her use of personal protective equipment. Such equipment may be external to the ear, in the form of an earmuff, or inserted into the outer ear, in the form of a plug. Such remedial measures must be accompanied by appropriate worker instruction and motivation and the clear marking of the noise zones where such protection is necessary.

Personal protection

If noise reduction measures do not achieve the desired standard, and further reduction is technically impractical or economically unrealistic, then there will be a need to resort to personal protection. This is obviously not ideal. It would be preferable to work in a quiet environment than to suffer the inconvenience, and sometimes discomfort, of hearing protectors. Nevertheless, properly used protection devices can achieve full protection against deafness in the majority of situations. And, if carefully selected, can result in only limited discomfort.

The types of hearing protection are usually described as earmuffs or earplugs.

Earmuffs

Earmuffs consist of high-attenuation cups which fit over the ears and are held firmly in position by a steel or plastic headband. The seal around the head is achieved with a soft cushion which, in the most efficient muffs, is filled with liquid. The slight improvement in efficiency obtained from liquid-filled seals must be offset against the fact that the liquid seal will leak if damaged. All seals should be easily replaceable.

The inside of an earmuff cup is partly filled with an absorbent material to reduce resonance in the shell.

It is possible to obtain earmuffs which incorporate a facility for electronic communication: music or messages are broadcast inside the earmuff cups. This is achieved by wire connections to fixed points, or by transmission through an induction loop so that no connections are needed. This is an expensive form of protection, but may overcome the isolation that some people feel when wearing earmuffs.

Earplugs

Earplugs are plugs of soft flexible material with high sound attenuation properties, which are pushed into the entrance of the ear canal. Many earplugs are disposable and are intended to be used only once. These are usually made from cylinders of soft plastic foam, or from glass wool. The plastic foam plugs are rolled up in the fingers before insertion, and they then slowly expand to fit tightly into the ear canal. Glass wool plugs may either be preformed, or come in wads of material which the user folds into a plug. Both are commonly provided from dispenser machines fitted on the wall of the workplace.

Non-disposable earplugs are made from soft plastic or rubber. They are preformed plugs designed to fit closely in the ear canal. However, because the human ear does not come in a standard size, such plugs generally have to be available in a number of fittings.

Selection

The hearing protector is a very simple device. It is a mechanism for insulating the ear rather than the noise source. However, there is an enormous number of types and brands available and the selection of the best model requires some care. There are two considerations: the protector must suit both the workplace and the user.

The workplace

It goes without saying that the protector must cut out sufficient workplace noise to ensure that the user is at no risk of hearing damage. Ideally, this would be achieved by the attenuation of a given hearing protector being specified in A-weighted decibels. Thus, in a noisy environment of 100dB(A), we could be sure that ear plugs with at least 20dB(A) attenuation would be satisfactory (since we can reasonably assume that 80dB(A) and below does negligible damage).

Unfortunately, specifying hearing protectors with a simple dB(A) rating is not possible. The reason is that all protection devices have different efficiencies at different frequencies. In other words, a particular ear muff might be very efficient at cutting out noise at a frequency of around 1000Hz, but less effective at frequencies which are much higher or much lower.

Thus, the reduction in dB(A) would be high if the noise was around 1000Hz, but lower if the noise was around, say, 100Hz.

To summarise, the degree of protection given by a particular set of hearing protectors will depend upon the frequency distribution of the workplace noise, and the particular attenuation characteristics of the protectors themselves.

Obtaining the effectiveness of a particular hearing protector at various frequencies is reasonably straightforward – the manufacturers of all reputable equipment will provide the data on request. This is likely to take the form of a graph of attenuation in decibels (not A-weighted) at varying frequencies. It is now necessary to assess whether these attenuation characteristics will give adequate protection in the area of concern. This requires a frequency analysis of the workplace noise. The effective new level (to the wearer of the protectors) is then obtained by subtracting the attenuation figure from the actual decibel level within each frequency band. This is illustrated on the graph (Fig 6). Whilst this exercise can be used to compare one hearing protector with another, it does not directly produce the effective reduction in dB(A) level and, if we are working to a dB(A) standard, this will be required. It is therefore usual to plot the figures on curves showing the A-weighted decibel levels. This will show both the effective dB(A) level to the user of the hearing protectors, as well as indicating at what frequency further attenuation will be necessary to obtain a still lower dB(A) figure.

Having compared the attenuation characteristics of various hearing protectors against the requirements of a particular workplace, there will almost certainly be a range of protectors which will be suitable.

In fact, for fairly broad band noise which is not too excessive, virtually all the hearing protectors on the market would probably be satisfactory. The choice can then be made on other factors such as availability, ease of maintenance, and cost.

Muffs or plugs?

One of the most debated issues in choosing hearing protectors is whether it is preferable to use earmuffs or earplugs. From

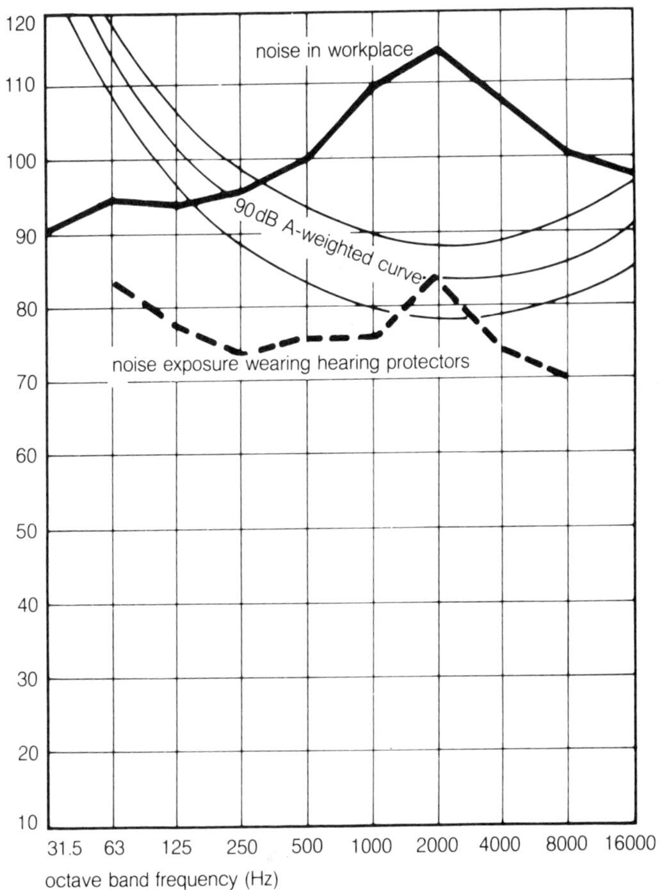

Fig 6 Graph illustrating the attenuation achieved by hearing protectors at different frequencies.

the point of view of protection, either can be perfectly adequate in the majority of circumstances. This is said with the slight reservation that some people find it difficult to use ear plugs. Perhaps this results from childhood conditioning not to push anything into the ears – whatever the reason it can mean that ear plugs are perched on the outside of the ears rather than firmly pushed into position. A further drawback of plugs, as against muffs, is that their use is less easy to supervise. Managers have responsibility and a legal duty to ensure that protective clothing is worn where necessary, and this is easier to achieve when the protection is reasonably conspicuous.

Offset against the above drawbacks of plugs are their lightness and convenience. They can be more easily stored and carried around the workplace than muffs. For many people they are more comfortable to wear than muffs because there is no pressure on the outside of the head and they do not make the ears hot.

For a large workforce, the deciding factor on the type of protection to be used may be the cost. The important requirement here is to consider the cost over an appropriate period of time. Disposable plugs at 30 pence might seem a better buy than muffs at £10. But if they are replaced every day, the plugs option will cost more than £70 per year.

Involving the workforce

Hearing protectors are amongst the more comfortable types of protective clothing for long term continuous use. However, no one would wear them from choice and an effective protection programme will require education, persuasion and assistance. The mere issue of protectors will not guarantee their use, and does not meet an employer's legal obligations. It is therefore essential that every employee who should be using hearing protectors is trained in the reasons why they are necessary, and the proper procedures for fitting them and keeping them in good order. Ideally, this exercise should include allowing a personal choice from a number of suitable alternatives.

Involvement of the workforce in the application of protective clothing is absolutely essential. However, it raises some questions which must be answered. The three main ones are:

(a) *Will hearing protectors prevent communication?*
To be heard in a noisy environment, it is necessary to shout over the background noise.
Hearing protectors will reduce both the background

noise and the shouting to the same extent and the ability to communicate remains the same. The answer is therefore that, when the noise level is over about 85 dB(A), hearing protectors do not impair the ability to communicate. However, this may psychologically be the impression, because wearing protectors creates an unusual environment for most people. The answer is to give the brain time to adjust. If the wearer can be persuaded to persevere for about two weeks, he or she will invariably find that their use becomes quite acceptable.

(b) *Can hearing protectors cause infections?*
The concern that protectors, particularly ear plugs, might cause infection, is common. Observations over a long period of time have shown that such a risk is negligible, and that the ear is quite resistant to repeated insertion of plugs. Nevertheless, hearing protectors must be kept scrupulously clean, and employees must have adequate facilities for proper cleaning and storage.

(c) *Surely I'm immune to noise by now?*
Employees who have already worked, perhaps for several years, in a noisy environment, can be particularly resistant to wearing hearing protectors. Their grounds are that noise can do them no further damage. It is essential to explain to such individuals the true nature of noise-induced deafness. No resistance builds up, and noise will continue to inflict damage until deafness becomes total.

Employees who have been exposed to noise, and who may already be partially deaf, have a vital need to preserve what is left of their hearing before it is too late.

Noise protection zones

A final point on the introduction of hearing protectors. It must be quite unambiguous where they should be worn. Areas in which protectors are to be used must be clearly marked; the boundary should be a conspicuous barrier or floor marking, and notices should make it obvious to everyone that protectors are required within the zone.

It is sometimes said that managers and visitors who are entering the zone for only a short time, need not wearing protectors. Whilst this may be correct in terms of the noise dose received, it is totally unworkable in practice. The only way to enforce a noise protection policy is for everyone to be bound by it at all times. The alternative is interminable discussions about how long an individual has been in the zone, etc. In a noise protection zone, everybody *must wear hearing protectors all of the time.* When the projected regulations take effect it will be obligatory.

Chapter 6

Management action plan

The objective of any management hearing protection policy may be achieved by a step-by-step plan. This will include consideration of the following: surveying workplace noise; setting a standard; reduction of noise by simple obvious measures; marking noise zones; choosing and enforcing the use of hearing protectors; specifying noise limits for new equipment; audiometry; checking and reviewing progress with the overall noise control policy; and ensuring the existence of adequate records.

Action checklist

This book has so far reviewed the legal and technical background to the subject of noise-induced deafness. However, this is now a preventable condition, and managers are charged with ensuring that the likelihood of it occurring in the workplace under their control, is negligible. This final chapter is presented as an action checklist which, if implemented, will essentially prevent noise-induced deafness. In addition, it will ensure that all legal obligations relating to noise, are fulfilled.

Undertake a noise survey

As with many other problems, the solution begins with measurement. In general, it is necessary to know the dB(A) level in each working position in the workplace. However, measurements are not needed from areas which are obviously quiet. Equally, quite a number of measurements may be taken in the noisy areas. This will give an idea of variations during the day, which machines are particularly noisy, etc.

It is possible to continue the measurement exercise in great detail, repeating the survey regularly, and extending the measurements to frequency analysis, and a number of variants on the decibel. Some of this may be worthwhile. However, it is common to see excessive measurement of an occupational health problem used as a substitute for doing something about it.

Further measurement is only necessary when additional information is required, or when there is a reason to believe that noise levels might have changed.

Determine noise dose

Identification of high dB(A) levels does not necessarily prove a noise hazard since there may be very limited exposure to those levels. Thus, if noise levels vary through the day, or, if employees are moving around, the noise dose of individuals must be obtained. The figure needed is the individual's continuous noise level over the working day - the Leq (8hr).

The Leq may be relatively easy to estimate on the basis that doubling the sound energy increases the dB(A) level by 3. Thus, if an individual spends half his or her working day (4 hours) at a noise level of 95dB(A), the 8 hour Leq will be 92dB(A), assuming that the rest of his or her time is spent in a quiet environment.

However, if the employee works at many different noise levels, the Leq becomes virtually impossible to calculate and must be measured. This is most conveniently achieved with a personal noise dose meter which the employee "wears" throughout the day. At the end of the day, the Leq can be read off directly.

Set a standard

The first two stages of this action plan will have produced some numbers, mostly dB(A) and Leq readings. It is now necessary to decide whether these findings are satisfactory, or whether action is required to reduce them. It is clear that the current legal situation in the United Kingdom requires that an Leq (8hrs) of 90dB(A) is set as the basic standard. However, we may wish to set a lower level as the standard for the following reasons:

(a) future legislation might be based on a lower action level;
(b) a lower standard leaves some margin for error in noise control and protection measures;
(c) it is desirable to ensure protection of a larger proportion of the workforce than is achieved by a 90dB(A) Leq.

Reduce all readily controlled noise

Before embarking on a complex and expensive programme of acoustic insulation, etc it is essential to ensure that all easily controlled noise is tackled.

This has been reviewed in Chapter 4, and includes elimination of machine guard rattling, steam leaks, etc.

It is at the end of such an exercise that the need for more sophisticated control measures should be reviewed. The legal requirement to reduce noise is not absolute; it is qualified by reasonable practicability. It is therefore acceptable to take cost and technical complexity into consideration in determining how far noise control measures should be taken.

If the actions taken so far have resulted in low noise levels, say less than 85dB(A) throughout the workplace, then no further action is required other than occasional checks that levels have not increased. However if, as is likely, some noisy activities are left, the remaining steps for hearing protection should be followed.

Establish hearing protection zones

A hearing protection exercise which requires employees to wear earmuffs or plugs must be unambiguous. In particular, there must be no doubt about exactly where protective clothing is required. Thus, the boundary of a protection zone must be clearly defined and marked with yellow lines. Notices should make it obvious that hearing protectors are required within the market area. Remember that the notices should comply with the Safety Signs and Colours Regulations and British Standard 5378. They should illustrate hearing protectors within a blue circle, and have the wording "hearing protection must be worn".

A point covered in Chapter 5 was whether to allow short-term access to the zone, without hearing protectors. In practice, this results in a breakdown of the discipline of using protectors, and should be resisted. It should be made clear to everyone, including senior managers, that protectors must be worn at all times within the zone.

At what noise level should a hearing protection zone be created? In setting the standard, it was suggested that an Leq (8hr) between 85 and 90dB(A) would be set. If the 90dB(A) Leq were selected, then noise zones should be created wherever the workplace noise exceeds 90dB(A), assuming that some employees will spend a full 8 hours in the noise zone. If employees spend less than 8 hours in the noise zone, there may be a temptation to increase the dB(A) level at which zones are created, on the grounds that the shorter exposure time will bring down the Leq level. In practice it is probably wiser to zone all areas where the level exceeds 90dB(A) knowing that then no one can possibly experience an Leq higher than this level.

that everyone understands the action the company will take when hearing protectors are not worn.

Companies vary in their disciplinary procedures, and there is no reason why the procedure used here should be any different from that used for other breaches of discipline. This is likely to entail informal and formal documented warnings, followed ultimately by removal from the job or dismissal. If the procedures are properly presented, and if it is clear that all management is subjected to the same rules, it is very unlikely that employees and their representatives will oppose such action.

In practice, of course, the involving of a formal disciplinary procedure should be a very rare event. Protective equipment enforcement is primarily a matter of effective management and supervision. Thus the training, job description, etc, of relevant management should include the importance of maintaining health and safety measures.

Ensure employee training is adequate

Many employees do not wear the hearing protectors provided for them. Often management attributes this to stubbornness or even stupidity. However, no one would voluntarily suffer avoidable deafness; it is entirely a problem of communication. It is absolutely essential that employees understand the risk of deafness, and that they appreciate the serious social disability from which they could suffer. It is also important that they know that noise-induced deafness is permanent and that a hearing aid will not help.

All of this calls for a good training programme. If the resource or ability does not exist within the company, then there are a number of independent trainers who can present the message most effectively. In addition, some good films are available which can at least supplement a course of training.

Training never ends, and re-training is necessary periodically to keep the message fresh. In general, employees in noise areas should be re-trained at least every two years, though relatively brief refresher sessions should be adequate for most circumstances.

The above comments have referred to training essentially as a motivation tool, to ensure the use of hearing protectors.

Check the adequacy of hearing protectors

It was explained in Chapter 5, that hearing protectors should be selected to suit the environment in which they are used. In other words, the protectors should be particularly efficient at any noise frequencies which are dominant in the workplace, and should reduce the effective dB(A) level by an adequate amount.

This is not to suggest that a major study involving frequency analysis and complicated calculation is required every time hearing protectors are selected. If the noise level is, say, 95dB(A) and clearly does not consist of "pure tone" frequency, then the vast majority of protectors are likely to be suitable and the selection can be made on the basis of comfort and cost.

For higher noise levels however, it is worth selecting the protectors more carefully. This involves subtracting the attenuation figures for the hearing protectors under consideration, from the frequency analysis of workplace noise. This gives the effective frequency breakdown for the protected individual. If this analysis is compared with the A-weighting curve, the effective new dB(A) level can be obtained.

If this exercise seems rather daunting, it may be possible to persuade the suppliers of the protectors to advise on the effectiveness of their models in a particular environment. If you wish to assess a wide range of protectors, it may be of interest to know that a number of computer programs exist which greatly simplify the process. For example, the Steel Castings Research Association has a program which compares the characteristics of a wide range of hearing protectors with a particular noisy environment, and identifies which protectors would be most suitable.

Establish effective enforcement of hearing protector use

By far the most common cause of breakdown in a hearing pro- tection programme, is the failure of employees to wear th protective equipment. In the main, this represents a commu cations problem and must be tackled by publicity and traini However, the fall back position should also be establishe

However, the training should also include the practical requirements of fitting hearing protectors, cleaning and maintaining protectors, using noise control items, etc.

Set noise specification for new equipment

The best laid plans for noise control and protection can be ruined by the arrival of a new machine. The workplace noise distribution is changed, noise zones must be redefined, and priorities for noise control are affected.

It is therefore necessary to specify the acceptable noise output from new machinery and to recognise that, if the specification is exceeded, there will be significant implications for hearing conservation. Unfortunately, the simple approach of merely defining a maximum dB(A) level for a new machine is no use in practice. This is equivalent to trying to buy a new fire by specifying the temperature it should produce. The dB(A) level is not a characteristic of a machine, and the noise level a new machine will produce depends to a large extent upon the surroundings in which it is placed.

Thus, whilst it is entirely appropriate to make a policy decision that no new machine should be introduced which takes the workplace noise level, say, over 90dB(A), it is not easy to predict whether a particular machine will have that result. It is necessary to ask the machine suppliers to specify what noise level the machine will produce at working positions in your particular workplace. If you wish to undertake the calculation in-house, then you will need to know the "sound power level" from the new machine and consult the acoustic textbooks in the bibliography to convert the power level to a dB(A) level for a specific point in a particular environment.

Review the response to audiogram measurements

There are many benefits in undertaking audiometric tests of employees who work in noisy areas. They can demonstrate the importance which management places upon the problem, and can identify deafness at a sufficiently early stage to take action which prevents serious disablement.

However, it is common to see a company introduce a programme of audiometry without recognising its implications. There is no point in obtaining information about the extent of employees' deafness unless you are going to do something as a result. Therefore, before any audiometric tests are undertaken, you must be able to answer the question "Exactly what will you do when you detect that an employee is partially deaf?"

Clearly, *some* response has to be made. To ignore the information would be irresponsible. An appropriate response could be to explain the findings to the employee, to retrain him or her in the use of hearing protectors, to increase supervision of the use of protectors, and to retest his or her hearing regularly. However, until the systems are in place to ensure that this can be done effectively, audiometry will serve little purpose.

Check policies and procedures

All companies with more than five employees must have a written statement of their safety policy.

This is required to include details of the arrangements for ensuring employees' health and safety. To be effective, the safety policy must be well written, frequently updated, and actively promoted by management. If noise is a particular workplace hazard, it is entirely appropriate to refer to this in the policy, and to state what control and protection measures are to be taken.

There are many other policy and procedure documents which may require revision to include a new approach to noise. For example, conditions of employment may have to refer to the requirement to use hearing protection. There might also be specific systems of work or permits, which should include the requirement for noise control or protection. Personnel and training procedures may be required to monitor the issue and instruction programme for hearing protectors. Finally, purchase procedures and specifications should be prepared so that no one can bring equipment into the workplace without first evaluating its impact on noise levels.

Check internal review mechanism

Once a company's approach to noise is fully resolved, it is

necessary to recognise the possibility that future changes in law or medical knowledge might require a change in policy. It might also be that changes become desirable in the light of experience. For example, following employees' reaction to a particular hearing protector.

It is therefore worth asking whether the company has a suitable mechanism for periodically reviewing the noise situation, and making changes where appropriate. A likely vehicle is a high-level committee, for example a company or group safety committee, which has overall policy making authority in this field. If noise is made a fixed item on their agenda for, say, an annual review, then actions on noise can be consistent and co-ordinated throughout the company, and the response to noise can be kept up to date.

Check documentation

Finally, having made an impeccable response to the problems of noise, put it in writing.

Records of noise surveys, issue of hearing protection, employee training, audiogram measurements, etc are essential. They are the basis from which the next level of actions will be taken. It must also be recognised that noise-induced deafness is an active area for legal proceedings, and documentation of actions may be required as evidence.

Good record keeping requires advance planning. Records which might seem adequate now, can be quite incomprehensible when retrieved in 10 years' time. The records need to be unambiguous, authoritative, and stored so that they can be readily retrieved. Consider your existing records and ask whether you can demonstrate what noise levels existed throughout the plant in 1975. Can you also show what hearing protectors a particular individual had been issued with, when they were last inspected, and when he or she was trained in their use? If answering these questions presents any problems, then the record keeping system would probably benefit from an overhaul.

Code of Practice for Reducing the Exposure of Employed Persons to Noise

HEALTH AND SAFETY EXECUTIVE

S. 1:
Scope of Code

1.1 *General application*

1.1.1 This Code of Practice applies to all persons employed in industry who are exposed to noise.

1.1.2 The Code sets out recommended limits to noise exposure. It should be noted that, on account of the large inherent variations of susceptibility between individuals, these limitations are not in themselves guaranteed to remove all risk of noise induced-hearing loss.

1.2 *Application to machinery*

1.2.1 This Code does not specify the measures which should be included at the design stage to control noise emitted by machines for which a separate code of practice is envisaged. A Technical Data Note (TDN 12) giving simple advice on measures which machine makers can incorporate in their product is available, free of charge, from any office of HM Factory Inspectorate.

1.3 *Relation to nuisance noise and vibration*

1.3.1 This Code does not seek to indicate measures for the reduction of community nuisance noise although it will clearly have an effect on this problem in some cases, nor does it include the effects of, or measures to be taken against, vibration.

S. 2:

Objectives of Code

2.1 *Specification of limits*

2.1.1 This Code specifies a limit for exposure to noise, and describes methods of measurement (Appendices 1 and 2) which can be used to determine whether the limit is exceeded.

2.1.2 Methods of measuring continuous steady noise, and for assessing a notional equivalent continuous sound level for fluctuating, intermittent, and impulsive noise are described (Appendix 3).

2.2 *Reduction of sound levels*

2.2.1 As a primary aim, the Code seeks the reduction of noise exposure to below the specified limit. As a secondary aim it seeks the reduction of sound levels generally.

2.2.2 The Code indicates appropriate measures for the reduction of noise exposure which should be taken by both management and employed persons.

S. 3:

Summary of measures to be taken

3.1 *Measures to be taken by management*

3.1.1 Management should accept a general responsibility for ensuring that the best practical means for noise reduction are applied.

3.1.2. The aim should be the general reduction of noise exposure. Where noise exposure less than the limits set in s. 4 are not achieved, ear protectors should be provided and their use ensured.

3.1.3 Appropriate staff should have adequate training in noise measurement and control.

3.1.4 Suitable records should be maintained.

3.1.5 Where it is not practical to ensure that the noise exposure is less than the limits set out in s. 4, and people must wear ear protectors, management should:

(a) identify and mark places where ear protectors are required,
(b) control entry into ear protection areas,
(c) ensure that suitable ear protectors are provided and are used,

(d) ensure that people provided with ear protectors are instructed in their care and use,
(e) where ear protectors are worn and the limit in s. 4 may still be exceeded at the wearer's ear, ensure that exposure periods are suitably reduced.

3.2 *Measures to be taken by employed persons*
3.2.1 Employed persons should:

(a) use and maintain measures adopted for noise control,
(b) report defective noise control equipment to the person responsible for its maintenance,
(c) use ear protectors when provided,
(d) not enter areas where ear protectors are required, unless authorised by management,
(e) not wilfully damage or misuse ear protectors provided and immediately report any damage or loss of such items to a responsible member of management.

S. 4:
Limits
4.1 *Desirable sound levels*
4.1.1 The limits set out in this section should be regarded as maximum acceptable levels and not as desirable levels. Where it is reasonably practicable to do so, it is desirable for the sound to be reduced to lower levels.

4.2 *Limiting sound level*
4.2.1 People should not be exposed to sound levels exceeding the limit set out in 4.3 to 4.5 below, unless they are using ear protectors which effectively reduce the sound level at the user's ear to or below the limits for unprotected ears. Note: The allowance for ear protectors, when worn, should be calculated as described in Appendices 4 and 5.

4.3 *Continuous exposure*
4.3.1 If exposure is continued for 8 hours in any one day, and is to a reasonably steady sound, the sound level should not exceed 90 dB(A).

4.4 *Non-continuous exposure*

4.4.1 If exposure is for a period other than 8 hours, or if the sound level is fluctuating, an equivalent continuous sound level (Leq) may be calculated and this value should not exceed 90 dB(A). Practical rules for calculating the value of equivalent continuous sound level from the readings of conventional instruments, and a mathematical definition, are given in Appendix 3.

4.4.2 Conventional sound level meters will indicate the value of Leq directly only when the sound level is continuous and reasonably steady. In other cases (fluctuating, intermittent or impulsive) various instruments, as described in Appendix 2 should be employed. For the purpose of this Code Leq should be considered to be adequately measured when these instruments have been used in the manner described. Alternatively any other instruments may be used provided the Leq is evaluated with adequate accuracy.

4.5 *Non-continuous exposure which cannot be adequately measured*

4.5.1 In certain circumstances, for example where employed persons move from one area to another, it may be difficult to measure and control exposure to non-continuous sound. If the non-continuous exposure cannot be adequately measured and controlled, any exposure at a sound level of 90 dB(A) or more should be regarded as exceeding the accepted limit and requiring the use of ear protectors. Places where this level is likely to be exceeded should be clearly identified (see s. 5).

4.6 *Sampling period for measurement of sound level*

4.6.1 When making measurements to determine whether the acceptable limit is exceeded it will not normally be necessary to measure the sound level during the entire working period. Assessment may be based upon sample periods which are typical of the working day.

4.7 *Overriding limits*

4.7.1 The A-weighted sound levels set out above are subject to an overriding condition that the unprotected ear should not be exposed to a sound pressure level, measured with an instrument set to the "fast" response, exceeding 135 dB, or in the case of impulsive noise an instantaneous sound pressure exceeding 150 dB.

4.7.2 Other parts of the body should not be exposed to a sound pressure level, measured with an instrument set to the "fast" response, exceeding 150 dB.

S. 5:
Surveys and identification of places where the sound level is excessive

5.1 *Surveys*

5.1.1 All places where it is considered the limit in s. 4 may be exceeded should be surveyed.

5.1.2 Surveys are advisable when it is necessary to shout in order to be audible to a person about one metre distant.

5.1.3 Places which are marginally below the limits set out in s. 4 should be re-surveyed whenever any changes are made which may alter the sound level.

5.1.4 Surveys should be carried out by a person who has received adequate training in noise measurement techniques.

5.2 *Identification of areas*

5.2.1 Areas where persons may be exposed to sound levels exceeding the limits set out in s. 4 should be identified as ear protection areas, and the boundaries clearly defined.

5.2.2 Entry to ear protection areas should be restricted to those authorised to do so. All such persons should use effective ear protection.

5.2.3 A prominent warning notice banning unauthorised entry, and entry without the use of ear protection should be posted near every entrance to an ear protection area. A suitable notice is illustrated in Appendix 6.

5.3 *Identification of machines*

5.3.1 Machines which, in normal operation, are likely to produce a sound level, at the operator's ear exceeding the limit in s. 4 should carry a prominent warning notice. The notice should be situated in a position clearly visible to the operator. A suitable notice is illustrated in Appendix 6.

S. 6:
Methods of controlling noise exposure

6.1 *General*

6.1.1 The best practical means for noise reduction should be

applied to all premises to which this Code applies.
6.1.2 The need for noise control should be taken into account when deciding which of different production methods or processes is to be used.
6.1.3 Reduction of noise is always desirable, whether or not it is practical to reduce the sound level to the limit set out in s. 4, and whether or not it is also necessary for people to use ear protectors. Reduction below the limit in s. 4 is desirable in order to reduce noise nuisance.
6.1.4 Some of the measures which can be taken to control the sound level are indicated in 6.2 to 6.6 below. This Code does not provide detailed technical information needed for carrying these measures into effect, or for deciding which measures are appropriate in particular circumstances. For this information reference should be made to text books and scientific literature.
6.1.5 It should be noted that when designing efficient and economic measures for noise control of existing machines the measurement of the sound produced in terms of A-weighted sound level [dB(A)] may need to be supplemented by some form of frequency analysis (eg octave band analysis). Advice on instruments for this purpose is given in Appendix 2.
6.1.6 It is recommended that design and construction of noise control measures should be supervised by a person skilled in noise control techniques.
6.1.7 When introducing noise control measures care should be taken to ensure that safety, and a satisfactory standard of other environmental factors (eg temperature and ventilation), are maintained.

6.2 *Separation of noisy areas*
6.2.1 Where practicable machines and processes producing sound levels in excess of the limit in s. 4 should be set apart.
6.2.2 Suitable partitions may be needed to prevent spread of noise. It is important that these be of correct size and location in relation to the size of the noise source, and the frequency of the sound to be intercepted.
6.2.3 Where a room, or building, is being divided into noisy and quiet areas, it is preferable for the separation to be made as complete as possible. This may be achieved by extending partitions to the walls, and the ceiling (except where false ceilings are installed) or roof, and by ensuring that there is a minimum of openings in the partition.

6.2.4 It may be advisable to provide sound-absorbing material in the noisy areas in order to prevent increase of sound level due to reflection from the walls and ceiling. The absorbing material should be such that a fire or health hazard is not introduced.

6.3 *Exhaust silencing*
6.3.1 Exhaust systems (including internal combustion engine and air exhausts) should be provided with effective silencers, or should be discharged in an area remote from employed persons, but not so as to create a hazard or nuisance to the public.
6.3.2 Silencers should be regularly inspected and maintained.

6.4 *Machine enclosure*
6.4.1 Where practicable noisy machines should be provided with sound-insulating enclosures. The operator should normally remain outside the enclosure.

6.5 *Enclosure of the operator's workplace*
6.5.1 In certain circumstances it may be possible and advisable to protect the machine operator by providing a sound-reducing enclosure or cabin.
6.5.2 When providing an enclosure for the operator due regard should be paid to the comfort of the occupant. In particular efficient ventilation and temperature control should be ensured. The enclosure should be as large as is reasonably practicable.
6.5.3 When enclosure of the operator's normal workplace is not considered appropriate it may be practical to provide a noise refuge which he can occupy when not actually working at the machine. A noise refuge is particularly suitable where an operator has no fixed position.

6.6 *Use of quiet machines and processes*
6.6.1 When deciding which of different production methods or processes is to be used the need for noise control should be taken into account as well as other factors. It may be possible to control noise by using a quiet process in place of a noisy one.
6.6.2 When appropriate, machines should be supported on anti-vibration mountings.
6.6.3 Metal to metal impact should be eliminated where possible.

6.7 *Inspection and maintenance*
6.7.1 All machines and areas where the sound level is above, or marginally below, the limit set out in s. 4 should be regularly inspected by a competent person for efficiency of the means for noise control. It is desirable that all new machines should be inspected after installation.
6.7.2 Persons employed on installation and maintenance should receive adequate training in methods of noise control. This should include instruction in the importance of correct installation, lubrication and adjustment of machines, and in the maintenance of any exhaust silencers which may be used as means of preventing excessive noise.

6.8 *Community noise nuisance*
6.8.1 Due regard should be paid to the need for avoiding the creation of noise nuisance in neighbouring property.

S. 7:
Ear protectors
7.1 *General*
7.1.1 When the application of means for controlling sound at source, or restriction of exposure duration, does not reduce the noise exposure to below the limit set out in s. 4, employed persons should be supplied with effective ear protection on an individual basis.
7.1.2 Ear protectors should not be used as a substitute for effective noise control. They should normally be regarded as an interim measure while control of noise exposure by other means is being perfected.

7.2 *Selection of ear protectors*
7.2.1 It is important to ensure that protectors will provide the majority of wearers with reliable and adequate protection. Suppliers should be instructed to provide full information on the sound reduction likely to be provided, the methods used for testing the protectors and details of the laboratory carrying out the test. Appendix 4 describes current test methods, and a procedure for using the test data for predicting the reduction of sound level.
7.2.2 Provided that adequate protection is given, it is preferable for the user to be allowed a personal choice among different patterns of protectors.

7.3 *Inspection and maintenance*
7.3.1 Ear protectors should be regularly inspected to make sure that they are undamaged and have not deteriorated.
7.3.2 Adequate provision should be made for clean storage of protectors when not in use.

7.4 *Education and joint consultation*
7.4.1 Before protectors are issued the need for their use should be fully explained and education should continue thereafter. Managers and supervisors should encourage the use of protectors by explanation and personal example.
7.4.2 Education of employed persons issued with ear protectors should include instruction in their use, care and maintenance.
7.4.3 Joint consultation before the issue of ear protectors is recommended.

S. 8:
Reduction of exposure duration
8.1 *General*
8.1.1 If it is necessary for people to work in places in which the sound level exceeds 90 dB(A) it may be possible to avoid excessive noise exposure of the people by reducing exposure duration.
8.1.2 Possible measures for restricting exposure duration include:

(a) job rotation,
(b) re-arranging of work to allow part to be carried out in a quiet place,
(c) where practical, arranging for jobs involving short-duration exposure to high sound levels to be performed by people who spend the rest of the day in quiet places, and not by people already exposed to noise near the limit in s. 4,
(d) provision of a noise refuge at the place of work.

8.2 *Rest rooms*
8.2.1 Where appropriate rest rooms should be separated from noisy areas by effective sound-insulating barrier.

S. 9:
New machinery

9.1 *Specification*

9.1.1 When purchasing new machinery the specification should require that the maker incorporates the best practical means for noise control (Appendix 7).

9.1.2 The seller should be required to furnish full information on the sound level likely to be produced. Current British Standard tests should be used where they are applicable and will indicate the sound level produced at the worker's ear, otherwise the test procedure should be agreed between the seller and the purchaser. The seller should be required to give details of the conditions of load during the test, and asked to indicate whether the test levels are likely to be typical of those produced with the machine in normal use in the user's factory.

9.1.3 It should be noted that the sound level produced in test conditions may be less than the sound level near to the same machine when installed in a factory. Possible reasons for an increase include reflection from the walls, floor and ceiling of the workroom, differences in mounting and loading conditions, and the additive effect of noise from nearby machines. It is, therefore, necessary to check the sound level after installation when there is any doubt as to whether or not the limit in s. 4 is exceeded.

S. 10:
Training

10.1 *Training of engineers*

10.1.1 Persons engaged in the specification, layout and installation of machine tools and factories should be adequately trained in techniques of noise measurement and control, or should be advised by suitably trained technical personnel.

10.2 *Training of others*

10.2.1 Other employed persons should receive such training as is necessary in the correct installation, operation and use of machines to avoid production of unnecessary noise. As appropriate this training should include instruction in the correct lubrication, adjustment, replacement of worn and loose or unbalanced parts of machines, and in the need for effective and correct maintenance of exhaust silencers and enclosures.

10.2.2 All persons exposed to noise should be acquainted with the hazards involved.

S. 11:

Records

11.1 *General*

11.1.1 Records of sound levels throughout the factory should be maintained to enable ear protection areas to be determined.

11.1.2 The records should include:

(a) the department and place in which readings were taken,
(b) the number of persons normally employed in the place concerned,
(c) the range of sound levels recorded,
(d) Leq if calculated,
(e) instruments used in the survey,
(g) date on which measurements were taken,
(h) any other relevant factor.

11.2 *Personal records*

11.2.1 Records of personal ear protection used should be maintained.

11.2.2 The records should include:

(a) the name of the person,
(b) job, department and place at which the person works,
(c) type of ear protection issued,
(d) date of issue of ear protection where appropriate.

Nomogram for calculation of equivalent continuous sound level (Leq) for an 8 hour period

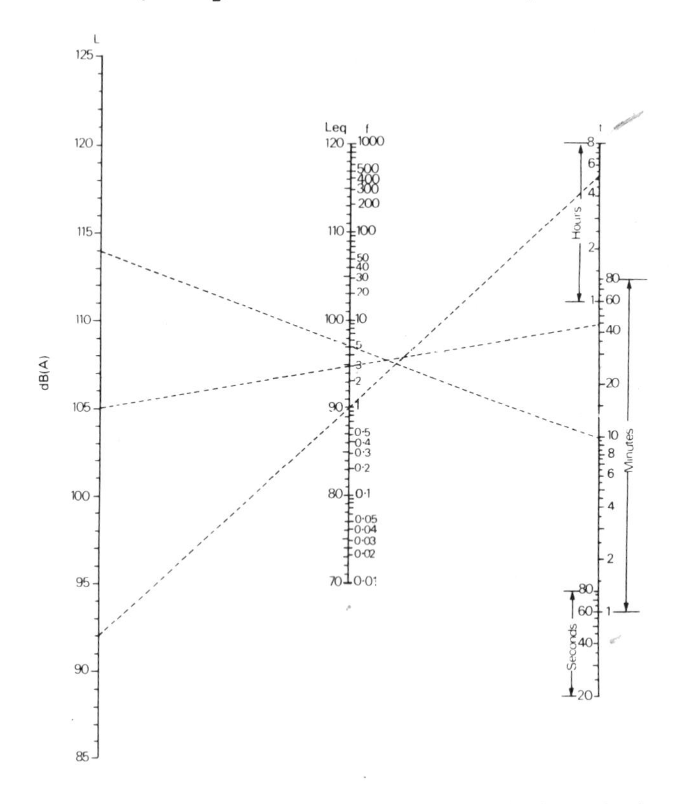

(1) For each exposure connect sound level dB(A) with exposure duration t. and read fractional exposure f. on centre scale.

(2) Add together values of f. received during one day to obtain total value of f.

(3) Read equivalent continuous sound level Leq. opposite total value of f.

Bibliography

100 practical applications for noise reduction methods. HSE/HMSO, 1983

Background to the HSC Consultative Document on the Protection of Hearing at Work – "Some aspects of noise and hearing loss: Notes on the problem of noise at work, and report of the HSE Working Group on Machinery Noise". HSC/HMSO, 1981
Note: Appendix D of this document itself contains a wide-ranging and comprehensive bibliography on noise and hearing protection, and also includes references to relevant British and International Standards on the subject.

CIS Bibliography No 14 Noise.
International Labour Organisation, 1979

Code of Practice for reducing the exposure of employed persons to Noise. HSE/HMSO, 1972, eighth impression 1978 (reproduced on pages 69–79)

Consultative Document "Prevention of damage to hearing from noise at work" – Draft proposals for Regulations and Guidance.
HSC/HMSO, 1987

Consultative Document "Protection of Hearing at Work" – Content of proposed Regulations and Draft Approved Code of Practice and Guidance Note. HSC/HMSO, 1981

Sharland I, Woods Practical Guide to Noise Control. Woods of Colchester, 1972

Taylor R Noise. Pelican, 1979 (third edition)

Webb J D (Editor) Noise Control in Industry, Sound Research Laboratories Limited, 1976 (second edition 1978)

Index